THE CRAZY K̶̶̶̶̶̶̶̶̶̶̶

ANOTHER HARE-BRAIN SCIENCE TALE

BY JOHN A. HONEYCUTT

ILLUSTRATIONS BY ANA NASTEVSKA

ART DIRECTION BY KRISTINA ILIEVSKA

PRODUCTION BY LAYNE PETERSEN

With Love To:
Arianna, Kaitlyn, and Sailor

Copyright John A. Honeycutt 2014. All rights reserved. ©

MY FAMILY

IS CRAZY.

WELL, ACTUALLY, WE ARE THE "KINETICS."
PEOPLE CALL US THE "CRAZY KINETICS."

I DON'T MIND.

I THINK IT'S FUNNY.

MY DAD'S NAME IS "SOUND"

MY MOM'S NAME IS "RADIANT."

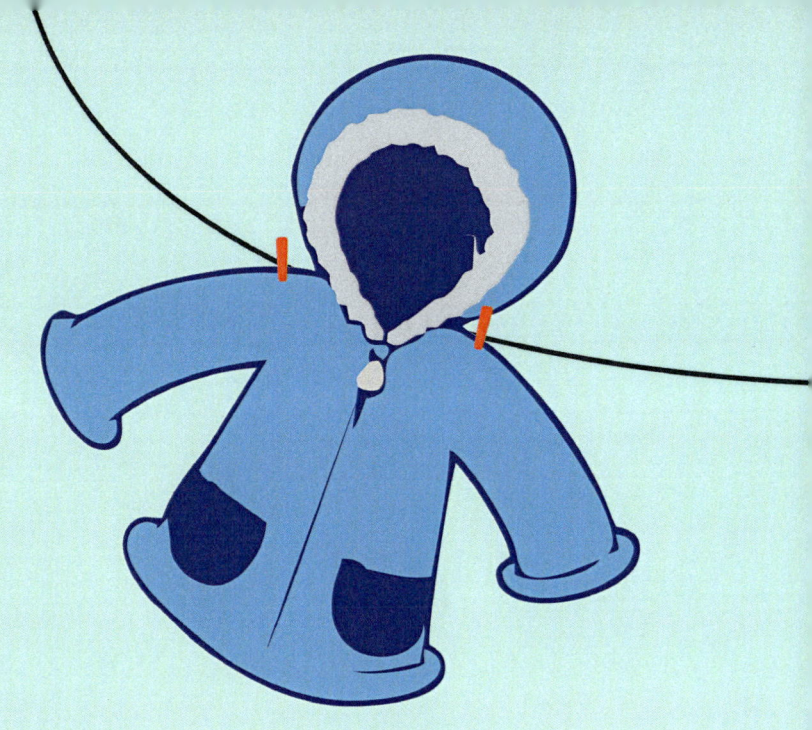

MY BROTHER'S NAME IS "THERMAL."

And I am "MOTION."

I LIKE TO MOVE AROUND A LOT.

WHEN DAD SAYS SOMETHING, YOU CAN DEFINITELY HEAR IT.

YESTERDAY HE SAID,
"MOTION, PLEASE TAKE OUT THE TRASH."
HE WAS IN A DIFFERENT ROOM.
BUT, I STILL HEARD HIM.
SOUND ENERGY CAN TRAVEL THROUGH WALLS.

ANYWAY, I GLADLY TOOK OUT THE TRASH.
I LIKE TO MOVE AROUND A LOT YOU KNOW.

MOM CAME IN TO HELP.
RADIANT ENERGY INCLUDES VISIBLE LIGHT.

Anyway, I found my basketball and played outside.

I LIKE TO MOVE
AROUND A LOT YOU KNOW.

MY BROTHER PLAYED TOO.

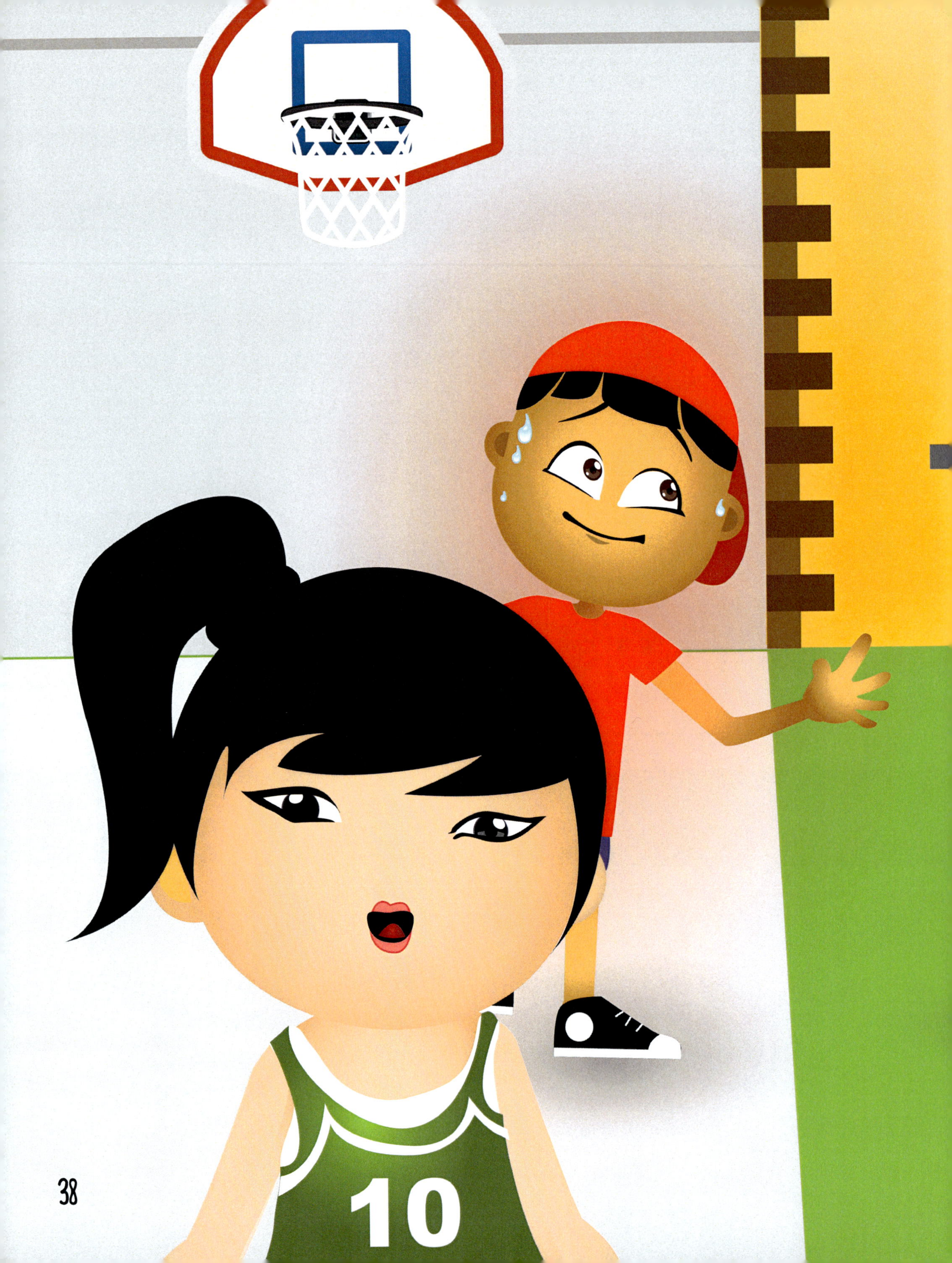
10

IT WAS REALLY HOT, BUT FUN TOO.
THERMAL ENERGY IS SOMETIMES CALLED HEAT.

Cereal
COFFEE
COFFEE
TEA
Beer
Beer
Beer
Beer
Spice
Tomato
SOUP
10

I WON'T TELL YOU WHO WON THE GAME.
YOU'LL HAVE TO GUESS.

42

BUT JUST REMEMBER, MY NAME IS "MOTION."

ANYWAY, YOU SHOULD VISIT US SOMETIME.

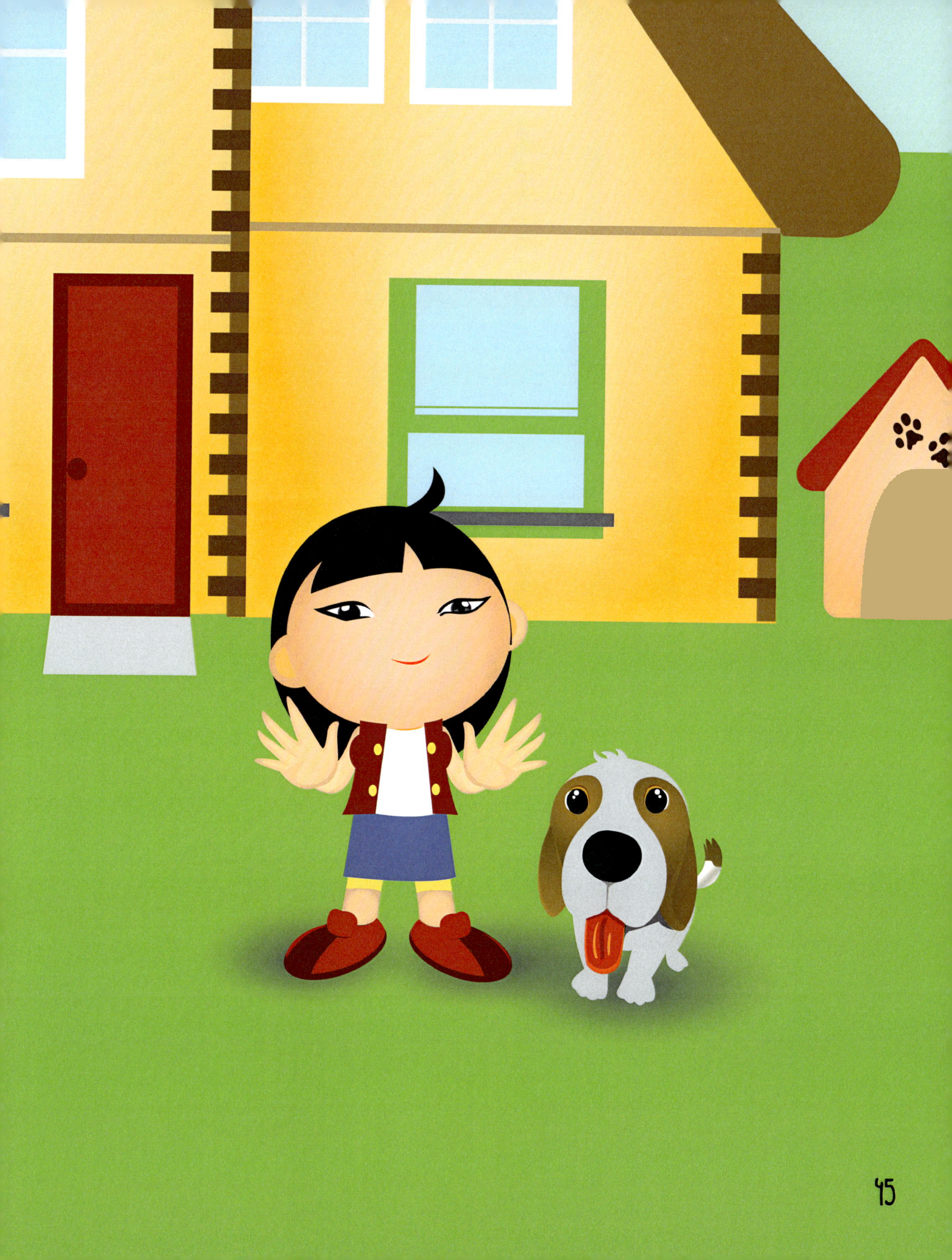

AND I WON'T MIND IF YOU CALL US THE "CRAZY KINETICS."

ALMOST ALL SCIENTISTS NEED TO KNOW ABOUT ENERGY. THE TOPIC OF ENERGY HAS A BUNCH OF VERY BIG WORDS — AND SOME OF THOSE WORDS ARE COMPLICATED.

BUT, SOME OF THE THINGS ARE NOT VERY COMPLICATED AT ALL. FOR EXAMPLE, THERE ARE ONLY TWO BIG CATEGORIES OF ENERGY: POTENTIAL ENERGY AND KINETIC ENERGY.

THIS BOOK INTRODUCES THE FOUR BASIC TYPES OF KINETIC ENERGY: MOTION, SOUND, LIGHT, AND HEAT.

MOTION — SOMETIMES CALLED MECHANICAL ENERGY — IS EASY TO IDENTIFY AND OBSERVE. WHEN THINGS ARE MOVING AROUND, THAT IS A FORM OF ENERGY IN ACTION.

SOUND — THE TERM TO DESCRIBE WHAT IS HEARD — IS MADE BY VIBRATIONS OF MOLECULES THROUGH WHICH THE SOUND TRAVELS. A GOOD

EXAMPLE OF THIS IS WHEN A DRUM OR CYMBAL IS STRUCK. DRUMS AND CYMBALS VIBRATE WHEN THEY ARE STRUCK. THESE VIBRATIONS MAKE AIR MOLECULES MOVE.

LIGHT IS RADIANT ENERGY. THE WORD "LIGHT" USUALLY REFERS TO ELECTRO-MAGNETIC RADIATION THAT IS VISIBLE TO THE HUMAN EYE.

HEAT IS ENERGY. SCIENTISTS USE THE WORD "THERMAL" TO MEAN "HEAT." HEAT ENERGY AND THERMAL ENERGY BOTH MEAN THE SAME THING. HEAT AND TEMPERATURE ARE NOT THE SAME THING. TEMPERATURE IS NOT ENERGY. TEMPERATURE IS JUST A MEASUREMENT OF ENERGY.

A FASCINATING FACT ABOUT ENERGY IS THAT ONE TYPE OF ENERGY CAN BE TRANSFORMED INTO ANOTHER TYPE OF ENERGY.

FOR MORE INFORMATION, VISIT HARE-BRAIN.COM.

Made in the USA
San Bernardino, CA
25 November 2014